MEGA MACHINES!

INCREDIBLE EARTHMOVERS

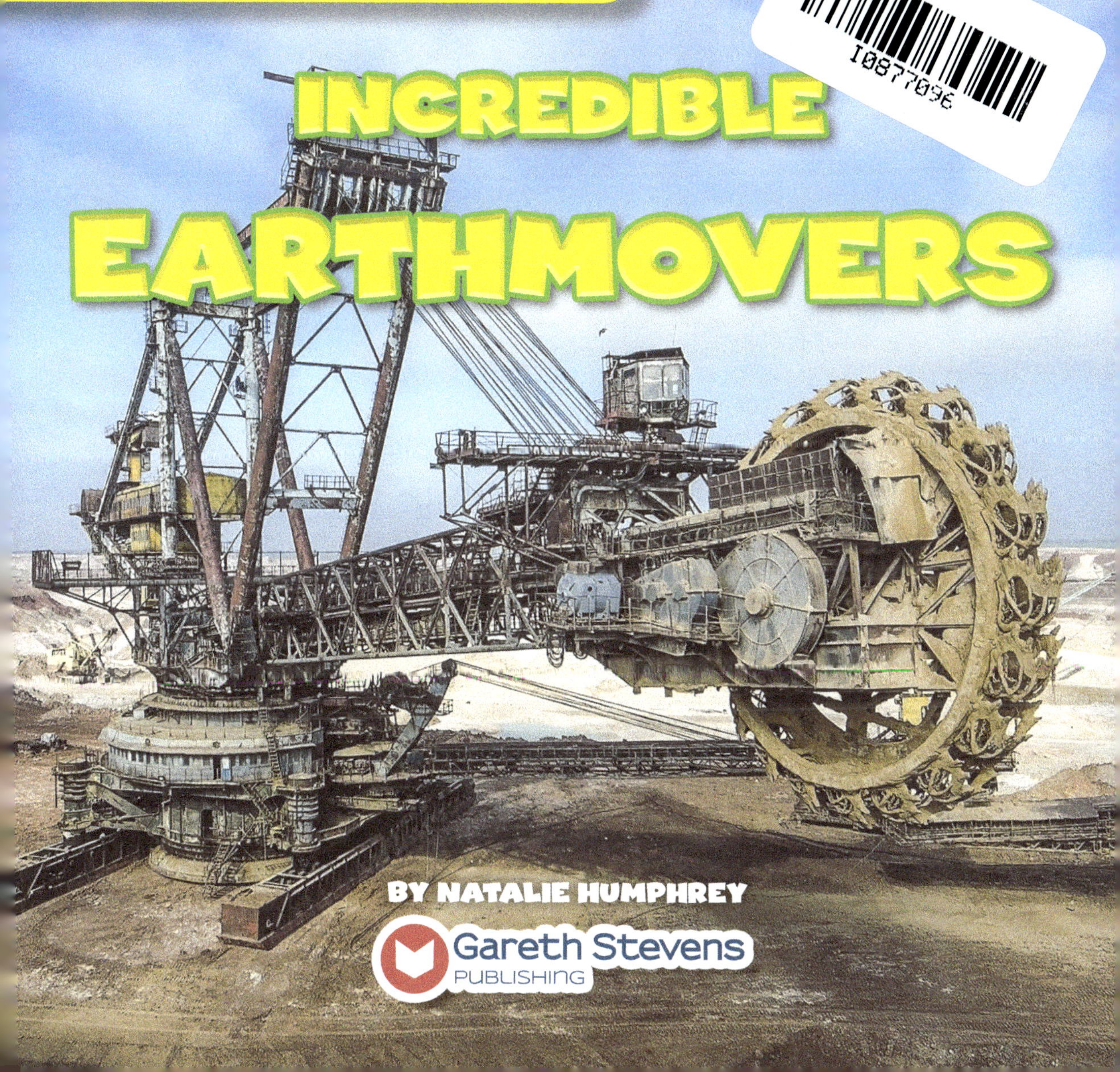

BY NATALIE HUMPHREY

Gareth Stevens PUBLISHING

Please visit our website, www.garethstevens.com. For a free color catalog of all our high-quality books, call toll free 1-800-542-2595 or fax 1-877-542-2596.

Portions of this work were originally authored by Kenny Allen and published as *Earth Movers*. All new material in this edition is authored by Natalie Humphrey.

Cataloging-in-Publication Data

Names: Humphrey, Natalie.
Title: Incredible earthmovers / Natalie Humphrey.
Description: New York : Gareth Stevens Publishing, 2023. | Series: Mega machines! | Includes glossary and index.
Identifiers: ISBN 9781538283110 (pbk.) | ISBN 9781538283134 (library bound) | ISBN 9781538283141 (ebook)
Subjects: LCSH: Earthmoving machinery–Juvenile literature. | Road machinery–Juvenile literature. | Earthwork–Juvenile literature.
Classification: LCC TA725.H86 2023 | DDC 621.8'65–dc23

Published in 2023 by
Gareth Stevens Publishing
2455 Clinton Street
Buffalo, NY 14224

Designer: Deanna Paternostro
Editor: Natalie Humphrey

Photo credits: Cover, p. Marianna Ianovska/Shutterstock.com; pp. 3, 4, 6, 8, 10, 12, 14, 16, 18, 20, 21 (bottom), 22, 23, 24 Nataliia K/Shutterstock.com; p. 5 Maksim Safaniuk/Shutterstock.com; p. 7 Smileus/Shutterstock.com; p. 9 mihalec/Shutterstock.com; p. 11 Mr. Tempter/Shutterstock.com; p. 13 F Armstrong Photography/Shutterstock.com; p. 15 Iuliia_K/Shutterstock.com; p. 17 g0d4ather/Shutterstock.com; p. 19 Andrea Slatter/Shutterstock.com; p. 21 (top) Aliaksandr Kushner/Shutterstock.com.

Printed in the United States of America

CPSIA compliance information: Batch #CW23GS: For further information contact Gareth Stevens at 1-800-542-2595.

CONTENTS

Boldface words appear in the glossary.

Mega Earthmovers

Earthmovers are a very important part of every dirty job. There are many different types of earthmovers. On **construction** sites, they help dig giant holes and get the ground ready for building. Earthmovers are also used in **mines** to help workers dig.

Bulldozers

One type of earthmover is a bulldozer. Bulldozers have a large front called a blade. The blade is used to push dirt and rock to make the ground flat. Instead of wheels, bulldozers use two **tracks** that allow them to move on soft dirt.

BLADE

Wheel Loaders

Wheel loaders look almost like bulldozers with some big differences. Wheel loaders have wheels to help them get around. They also have a big **scoop** instead of a flat blade. Wheel loaders can scoop up dirt, rocks, sand, and even snow!

SCOOP

Dump Trucks

When a wheel loader picks up a load, it can dump it into a dump truck. Dump trucks are used to **haul** loads to and from work sites. The box on the back of a dump truck tips up to dump out its load.

Scrapers and Graders

To build roads for cars to drive on, workers use scrapers and graders. Scrapers scoop up rock and dirt that gets left behind by bulldozers. This rock and dirt is put into a **hopper**. Then, graders use a long blade to make the ground totally flat.

HOPPER

Excavators

Excavators have many different dirty jobs. They have a long, moveable arm with a **bucket** at the end that can scoop dirt or knock things over. Excavators are used to dig holes, move heavy things, and are even used in **demolition**!

BUCKET

The TLB

The TLB, or tractor loader backhoe, is another type of earthmover. This machine is usually used for smaller construction jobs, building roads, or farming. The TLB is more **agile** than some bigger mega machines, but they can do jobs bigger machines can't!

TEREX
TEREX

Trenchers

To dig bigger holes, workers may use trenchers. Trenchers are used to dig holes for pipes and cables on construction sites. Trenchers don't always look the same. Many have giant saws or scoops that cut through dirt or even rock. Bigger trenchers have both!

Belaz 75710

Many dump trucks might be big, but the Belaz 75710 is the biggest dump truck in the world. Workers use this mega machine to carry loads of dirt from deep mines in the ground. This dump truck is taller than a two-story house!

Mega Machine Facts: the Belaz 75710

Weight: 360 tons (326.6 mt)

Height: 26 feet (8 m)

Max Load Weight: 496 tons (450 mt)

Top Speed: 42 miles (68 km) per hour

GLOSSARY

agile: Able to move quickly and easily.

bucket: A large container that many machines have for carrying dirt and other materials

construction: Having to do with the act of building something.

demolition: The act of purposefully demolishing, or tearing down, a building or other structure.

haul: To pull or carry something heavy.

hopper: A container that pours materials into another machine or bin.

mine: A pit or tunnel from which minerals are taken.

scoop: To reach under something and lift it.

track: A metal loop with wheels inside that moves some big machines.

FOR MORE INFORMATION

BOOKS

Honders, Christine. *Construction Workers.* New York, NY: PowerKids Press, 2020.

Loy, Harriet. *A Bulldozer's Day.* Minneapolis, MN: Bellwether Media, Inc.,2023.

WEBSITES

Britannica Kids: Mining
kids.britannica.com/kids/article/mining/353469
Learn more about how mines are made and how earthmovers help get the job done.

Wonderopolis
wonderopolis.org/wonder/how-big-is-a-dump-truck
Learn more about the biggest dump trucks and other jobs they're used for.

Publisher's note to educators and parents: Our editors have carefully reviewed these websites to ensure that they are suitable for students. Many websites change frequently, however, and we cannot guarantee that a site's future contents will continue to meet our high standards of quality and educational value. Be advised that students should be closely supervised whenever they access the internet.

INDEX